THE YACHTSMAN'S
WEATHER MANUAL

THE YACHTSMAN'S WEATHER MANUAL

Jim McCollam

Illustrated with diagrams by the author

DODD, MEAD & COMPANY New York

ISBN: 0-396-06721-2
Library of Congress Catalog Card Number: 72-7198
Printed in the United States of America
by Vail-Ballou Press, Inc., Binghamton, N. Y.

To BILL ROBINSON, Editor of *Yachting,* whose encouragement made this possible

INTRODUCTION

WHILE everyone talks about the weather, yachtsmen are more affected by it, with the possible exception of aviators, than anyone else.

The cruising or day sailing yachtsman needs to know about impending storms in order either to find a snug harbor before they hit or to prepare for them. If making a passage, he can plan better when shifts in direction or velocity of wind can be forecast.

The racing yachtsman is even more dependent on weather. Races are won by knowing what's going to happen to the wind long before it does happen and by sailing a course to best take advantage of a change in direction or velocity. Say you are on a reaching leg of a race. If you expect the wind to increase, it will pay of course to sail high in anticipation of bearing off later at high speed when the wind increases. The ability to forecast changes in wind direction is even more important to the racing sailor.

These are just some of the reasons why Jim McCollam's book, *The Yachtsman's Weather Manual*, is so important. The author makes no claim that by reading this book you will always be able to forecast correctly. Even the best of weather services muff their forecasts a certain percentage of the time. McCollam's aim is to develop in the reader an understanding of weather phenomena so that by knowing what *is* happening he will have a better chance to tell what *will* happen. He points out, however, that "weather forecasting is based on the science of the laws of probability, not a set of inviolate postulates like Euclid's geometry." In short, one can be wrong, and as if to prove this point, McCollam gave me a faulty forecast for the winds in the 1972 St. Petersburg to Venice Race. But he *did* mention that conditions were so unsettled that a definite forecast was risky and hence I had better keep my eyes open. The following week he vindicated himself by giving a perfect forecast for the St. Pete-Lauderdale Race.

The importance of knowing the weather was brought home to me most forcefully when I was navigator of Ted Hood's *Robin* when we won the 1968 Bermuda Race. We had on board a meteorologist, Joe Chase, who was almost uncanny in predicting what the wind would do. In asking Joe how he did it, I found it all more than a bit confusing and overly involved. I assumed that merely confirmed that I was stupid, but Jim McCollam's book has restored some degree of self-confidence. He doesn't try to give the reader a technical knowledge of meteorology but instead presents the subject in such simple yet authoritative terms that a better understanding of weather phenomena is inevitable, and with this improved understanding comes a better chance of accurate forecasts. I'm sure I shall still make errors in my forecasting, but even more sure that having read this book I shall be right more often than heretofore.

ROBERT N. BAVIER, JR.

CONTENTS

THE YACHTSMAN'S
WEATHER MANUAL

THE WORLD OF
WIND AND WATER

THE lure of the unknown led man to the water's edge. His imagination and creativity built the first crude boat. Necessity carved a paddle, but the sail was sewn by dreams of distant shores; haste conceived an engine.

The relationship of a man to his boat is intimate and personal; man must participate; he must not just control speed and direction, he must contribute intelligence to the driving force.

Without man a boat is an inanimate thing, drifting aimlessly in the variable and fickle winds. Let him take the helm and it becomes a thing of life and beauty.

We are used to an environment of straight lines and flat surfaces—horizontal and vertical. In the world of wind and water, we find the complex, compound curves of billowing sails, a careening deck. We must adjust our perpendicular equilibrium to a brand new geometry. Here are new and different sensations—sights and sounds to be found nowhere else. There is no better way to shed the burdens of an overpopulated, overcivilized world.

At night, enshrouded in a canopy of darkness, the stars are closer and more numerous than on the shores that have been left behind. It is easy to forget that there is another world with too much of everything but stars.

The moon that we take for granted on land becomes our companion of the night; we follow her progress with a diligence and fascination primeval.

At sea, man can only feel humility in the enormity of his dual environment of wind and water. Out of sight of land, all distance is infinite; the horizon is unapproachable; space is limitless, and the sea beneath him is unfathomable.

The world of wind and water is a coalition of two empires of such immensity and complexity that it staggers the mind. The power of the forces generated by each exceeds all the man-made energy in history, all the power produced since the invention of the wheel—not over the same period, but in a matter of hours. A major hurricane will release enough energy in a few days to supply the entire world with power for several years.

It is this world of wind and water that we intend to explore. We will not become meteorologists or oceanographers, but we will better understand the forces of nature with which we have to cope as yachtsmen.

Will it rain tomorrow? Will the wind blow? Will the clouds obscure the sun? Will the fog move in? The possibility of predicting these conditions with the infallibility of Pythagoras's theorem is slight.

If this is true, then what is the use of trying to learn anything about the weather? Well, isn't it worth knowing if there is a possibility of foul weather tomorrow? Isn't it worth knowing if the wind might blow this weekend? Wouldn't you like to know the probable direction and velocity?

Even if you are wrong part of the time, there is a comfortable feeling inherent in the understanding of the elements, just as there is a superstitious fear of the unknown.

There is a definite pattern to our weather system. But the fact that it is not easily forecast has been the butt of humorists for over a century. This humor is based on a lack of understanding of the problems involved.

Weather is the result of the heating and cooling of our atmosphere and the fluctuation of its moisture content. The factors that affect these changes are infinitely variable, overlapping and interlocking. Weather forecasting is based on the laws of probability, not a set of inviolate postulates like Euclid's geometry.

The fundamental things that affect our atmosphere and cause the endless variety and dramatic extremes of weather are our own planet, Earth, and the sun. This fiery sphere in space annually pours onto the earth 23,-000,000,000,000 (twenty-three trillion in the United States) horse power in electromagnetic waves; the earth's contribution is its rotation on an axis tilted at $23\frac{1}{2}°$.

Whirling in orbit at 66,000 mph, the earth rotates at 1040 mph. The canted axis results in the seasonal variables in temperature and the ratio of darkness to light. This spinning, wobbling, elliptical path mixes and circulates the atmosphere in a surprisingly uniform pattern. Even with this thorough mixing of cold and warm air, the waters of our lakes, rivers, and oceans are indispensable.

Assuming that there were inexhaustible subterranean wells to supply the water necessary to support life, without surface water the temperature extremes would be so great, survival would be impossible. The air would be in constant motion at velocities greater than those of the most savage hurricane. The temperature at the equator would continually rise higher and at the poles it would drop to absolute zero. The ultimate fate of the earth would be sheer speculation.

Instead, by a huge and complicated system of airflow or planetary winds, the tropics are cooled by the poles, which are reciprocally warmed by the tropics. Since dry air is a very poor heat transfer agent, it is the moisture carried poleward to condense and release latent heat that makes our survival possible.

On a minor scale, the circulation of the atmosphere is augmented by ocean currents that carry warm tropical water to the poles, and by frigid polar currents that flow toward the equator.

However, the major burden of balancing the temperature of the earth is carried by the thin gaseous film that envelopes the planet. Within the atmosphere the principal agents of heat transfer are the great planetary winds —the polar easterlies, the prevailing westerlies, and the trade winds.

Characteristic of the westerlies are the extratropical low pressure systems—the huge whirling air masses of the temperate zones.

It is beyond the scope of this book to cover the entire subject of meteorology; the goal is to promote a better understanding of weather in the western hemisphere north of the equator. It is directed specifically to yachting interests in that it will place more emphasis on wind than on the various subtle climatic variations. Since the state of the surface conditions of various bodies of water is germane to boating interests, waves, tides, and currents will also be examined.

You don't have to be a synoptic meteorologist to be a good sailor, but a superficial knowledge of weather phenomena is invaluable.

One of the things with which we shall be concerned is air pressure, which exerts 14.7 pounds per square inch at sea level. This force on the surface of mercury in a reservoir will support a column 29.92″ in a closed glass tube. The equivalent value in millibars is 1013.2. This unit has largely superseded inches of mercury because a millibar represents an actual force—1000 dynes; a dyne is the amount of force required to accelerate one gram, at one centimeter per second per second. (*See* Appendix.)

It should be pointed out that the standard atmospheric pressure is a hypothetical situation. The mean pressure at the island of Bermuda ranges from 1017 millibars in January to 1022 millibars in July, and that is considered normal. Conversely, the mean pressure in the Aleutians ranges from 1000 millibars in January to 1012 millibars in July.

The use of knots and sometimes miles per hour for giving a numerical value to the velocity of the wind is common practice. However, since it is impractical to relate these unit increments to the state of the sea and wave heights, the use of the Beaufort Scale by mariners is preferable.

In 1806 Sir Francis Beaufort, admiral of the British Navy, devised a scale of wind forces to categorize its velocity. With a range of 0 to 12 he provided a system of estimating and logging wind speed long before the advent of the anemometer. It was such a practical method that, in one form or another, it survives in common usage today and will be consistently used in this text.

The following is a composite wind velocity scale based on the original Beaufort system and that used by the U.S. Weather Bureau today.

Under the heading, "Wind Effects," Beaufort indicates the original; "At Sea" and "On Land" are the ones used currently.

Beaufort Number	Term	Knots	Wind Effects	Probable wave height
0	Calm	Less than one.	*Beaufort:* Calm. *At Sea:* Sea Calm and mirrorlike. *On Land:* Calm, smoke rises vertically.	0
1	Light Breeze	1–3	*Beaufort:* Just sufficient to give steerageway. *At Sea:* Scalelike ripples without foam crests. *On Land:* Smoke drifts.	3 in.
2	Light Breeze	4–6	*Beaufort:* Ship driven at 1–2 knots, full and by. *At Sea:* Small short wavelets; crests have glassy appearance and do not break. *On Land:* Wind felt on face; leaves rustle; ordinary vane moved by wind.	6 in.

Beaufort Number	Term	Knots	Wind Effects	Probable wave height
3	Gentle Breeze	7–10	*Beaufort:* Ship driven 3–4 knots, full and by. *At Sea:* Large wavelets; some crests break; foam of glassy appearance. Occasional white foam crests. *On Land:* Leaves and small twigs in constant motion; wind extends light flag.	2 ft.
4	Moderate Breeze	11–16	*Beaufort:* Ship driven at 5–6 knots, full and by. *At Sea:* Small waves, becoming longer; fairly frequent white foam crests. *On Land:* Raises dust, loose paper; small branches are moved.	4 ft.
5	Fresh Breeze	17–21	*Beaufort:* All clean sail, full and by. *At Sea:* Moderate waves, taking a more pronounced long form; many white foam crests; there may be some spray. *On Land:* Small trees in leaf begin to sway; crested wavelets form on inland lakes.	6 ft.
6	Strong Breeze	22–27	*Beaufort:* Single reefed topsails or topgallant sails. *At Sea:* Large waves begin to form; white foam crests everywhere; there may be some spray.	10 ft.

Beaufort Number	Term	Knots	Wind Effects	Probable wave height
			On Land: Large branches in motion; whistling heard in wires; umbrella use difficult.	
7	Near Gale	28–33	*Beaufort:* Double reefed topsails and jib. *At Sea:* Sea heaps up and white foam from waves begins to blow in streaks along direction of wind; spindrift begins. *On Land:* Whole trees in motion; hard to walk against the wind.	14 ft.
8	Gale	34–40	*Beaufort:* Triple-reefed topsails. *At Sea:* Moderately high waves of greater length; edges of crests break into spindrift; foam blows along the direction of wind. *On Land:* Breaks twigs off trees; generally impedes progress.	18 ft.
9	Strong Gale	41–47	*Beaufort:* Close-reefed topsails and courses. *At Sea:* Heavy waves; dense streaks of foam along the direction of wind; crests of waves begin to topple, tumble, and roll over; spray may reduce visibility. *On Land:* Slight structural damage occurs.	23 ft.

Beaufort Number	Term	Knots	Wind Effects	Probable wave height
10	Storm	48–55	*Beaufort:* Close-reefed main topsail and reefed foresail. *At Sea:* Very high waves with long overhanging crests. Foam in great patches is blown in dense white streaks along the direction of wind. The surface of the sea is white in appearance. The tumbling of the seas becomes heavy and shocklike; visibility reduced. *On Land:* Seldom experienced inland; trees uprooted; considerable structural damage.	29 ft.
11	Violent Storm	56–63	*Beaufort:* Storm staysails. *At Sea:* Exceptionally high waves that might obscure small- and medium-sized ships. The sea is completely covered with long white patches of foam lying along the direction of the wind. Everywhere the edges of the wave crests are blown into froth; visibility reduced. *On Land:* Very rarely experienced; accompanied by widespread damage.	37 ft.
12	Hurricane	64 or more	*Beaufort:* Bare poles; that which no canvas could withstand.	45 ft.

Beaufort Number	Term	Knots	Wind Effects	Probable wave height
			At Sea: Air filled with foam and spray; sea completely white with driving spray; visibility very much reduced.	
			On Land: Very rarely experienced; accompanied by widespread damage.	

ONE

PLANETARY WINDS

IF the earth were a king-size bowling ball, our weather would be quite simple and uniform but we wouldn't be here.

This planet of ours is land and water; its mountains are high and its seas are deep; it revolves about the sun on its wobbling, tilted axis; it heats and cools the atmosphere, sending it whirling about, ascending and descending.

Heated air expands, becomes less dense and rises; cooled air contracts, becomes more dense and sinks. Because of the higher pressure (density) of cold air, it flows out along the surface to the lower pressure areas of warm air. And so the winds blow.

The pressure difference or gradient is the motivating force; it also determines direction—from high to low pressure. In the polar regions, the cold heavy air flows along the surface toward the equator. In the tropics, the hot air ascends and streams poleward.

A few weeks before, the balmy southeasterly breeze across the Gulf of Mexico was sub-zero air sinking to the surface in the Arctic. The tropical zephyrs of Tahiti do

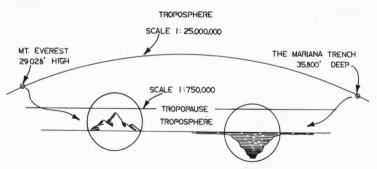

Fig. 1–1 The curved line drawn through Mt. Everest and the Mariana Trench does not represent the earth's surface; it is the depth of the troposphere (the part of our atmosphere where all weather takes place) at the scale of 1 to 25,000,000! Mt. Everest, highest point in the world, reaches to within a few thousand feet of the top of the troposphere. The Mariana Trench near the Phillipines is the deepest hole in the oceans. Even with the seas drained dry, the surface of the earth is relatively as smooth as the skin of an orange. It is difficult to conceive that the tremendous forces generated by air and water take place in a zone that is proportionately equal to the fuzz on a peach.

not stay there; they ultimately become the icy gale of the Siberian Steppes.

Before we can cope with the wanton and wayward winds of the waters upon which we sail, we must develop, not meteorological expertise, but a functional understanding of the basic circulation of the atmosphere.

This basic or general circulation is referred to as the primary, terrestrial, or planetary winds. Since the latter seems more explicit, it is the term that we will use.

For those of us whose academic days in halls of ivy are more remote than we care to admit, a little review of elementary physical science might be in order; perhaps the "now" generation will bear with us.

Since the heating and cooling of the atmosphere is germane, let us take a look at Fig. 1–2.

If there were three identical windows in space allowing an equal amount of the sun's rays to penetrate the atmosphere in various parts of the world, there would be considerable inequity in the distribution of heat. In the tropics, during March, the rays would be almost perpendicular and cover an area about equal to the size of the window. Because of the angle of incidence, the penetration of the atmosphere would be maximum, and the mean temperature at Singapore (1°17′N) would be about 81° F., pressure—1009 millibars.

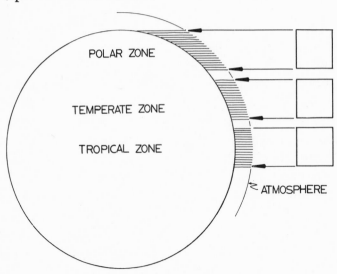

Fig. 1–2 The irregular heating of the earth's surface by the sun is the fundamental cause of weather phenomena. The most significant cause of this irregularity is the angle of incidence at which the sun's rays strike, not only the surface of the earth, but also the atmosphere. The air screens out some of the rays, reflects and refracts others back into space. The above diagram shows the manner in which the sun's rays are distributed in the three major climactic zones.

In the temperate zone, the rays would be spread out over about 25% more of the earth's surface, distributing somewhat fewer BTUs per square mile. In passing through the atmosphere, the rays would have to travel farther and in the process lose heat to the air molecules, water vapor, and dust particles. The angle of incidence would be more acute and more rays would be refracted and reflected back into space.

At Jacksonville, Florida (30°20′N), the mean temperature would be 63° F., pressure—1019 millibars.

In the polar zone, the same amount of heat is spread out over more than twice as many square miles as at the equator, and the angle of incidence is so acute that most of the rays bounce back off into space. As a result of this inefficient heating, the mean temperature at Spitzbergen (78°02′N) within the Arctic Circle is −5° F., pressure—1009 millibars.

As a matter of fact, the polar zone radiates more heat into space than it receives from the sun and the tropics receive more heat than they radiate. If it weren't for the circulation of the atmosphere, the polar zones would gradually drop to absolute zero and the temperature at the equator would be purely a matter of speculation.

Now everybody knows that it is hot at the equator and cold at the poles and maybe everybody knows why. So what's the fuss? Well, take a look at Fig. 1–3 and we'll see.

The air over the polar zone is chilled by the low surface temperature; it contracts, becomes more dense and therefore sinks; it flows out along the surface away from the pole. As it travels over terrain that has progressively

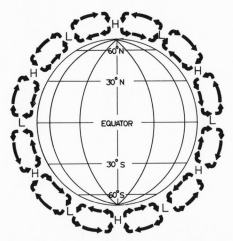

Fig. 1–3 This is the basic circulation of the atmosphere showing where rising air and low pressure bands occur and where sinking air and high pressure bands occur. Converging, rising air is always associated with low pressure and descending, diverging air is typical of high pressure conditions.

higher temperature, the air is warmed; it expands, becomes less dense and therefore lighter.

At about 60° latitude, it becomes so buoyant it starts to rise. It ascends to about 35,000 feet, becomes stable, and flows out to the north and south.

The air in the equatorial regions becomes heated and buoyant; it ascends to about 50,000 feet, becomes stable, and spreads out to the north and south.

This leaves upper level air flowing from 60° to the pole, where it becomes cold and sinks to the surface. The air of tropical origin at 50,000 feet becomes cooled and descends at about 30° latitude.

Well, what is so complicated about that? The wind blows either north or south or up or down and the pressure is high here and low there.

The problem is that we have a trouble maker—the rotation of the earth; it causes all our weather problems, like hurricanes and their ilk.

Let's say that Paul Bunyan and King Kong decided to toss a baseball back and forth between Key West, Florida, and the Mississippi Delta—a distance of about five hundred miles (Fig. 1–4).

If Paul stood on the Delta and heaved the horsehide at a leisurely 100 mph toward Kong, leaning against a palm

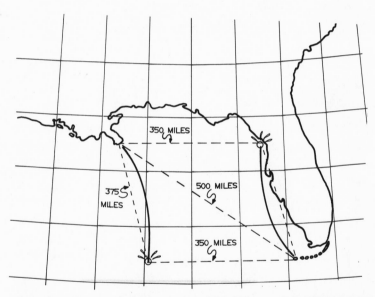

Fig. 1–4 Next to the angle of incidence of the sun's rays, the most important influence on our weather is the Coriolis Effect. As shown above, an object directed at the Mississippi Delta from Key West and traveling at 100 mph would fall far to the east in the Gulf of Mexico. Likewise, on the opposite course it would also be deflected to the right and miss its mark by 350 miles. (See text.)

tree on Key West, he would miss his target by a country mile—350 of them, to be more precise. On the other hand, Kong's toss would splash down offshore from Cedar Key, Florida, for no better performance than that of his partner.

The element responsible for all this wild pitching is the rotational velocity of the earth at various latitudes, which is:

Equator . . . .*1040 mph*. . . . .*mouth of Amazon River*
15°N . . . . . .*1000 mph*. . . . .*Martinique Island*
30°N . . . . . . *900 mph*. . . . .*Jacksonville, Fla.*
45°N . . . . . . *740 mph*. . . . .*Halifax, Nova Scotia*
60°N . . . . . . *520 mph*. . . . .*Anchorage, Alaska*
75°N . . . . . . *270 mph*. . . . .*Bylot Island, Baffin Bay*
90°N . . . . *0 mph*. . . . .*North Pole*

At Key West the surface velocity of the earth is about 70 mph faster than it is at the Delta. That means Key West was moving out from under Paul's pitch at that rate. When King Kong made his try, he was assisted by the 70 mph plus of his mound. Since Paul didn't know anything about the Coriolis Effect, he didn't try to run around the Gulf coast to field the toss.

Without becoming too technical, let it suffice to say that the Coriolis Effect causes anything moving over the earth's surface to be deflected to the right in the northern hemisphere and to the left in the southern hemisphere. It is called an "effect" because the actual path in relation to space is straight; it is the apparent path that is curved because of the rotation of the earth.

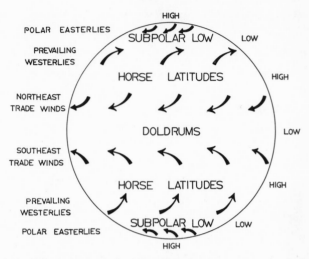

Fig. 1–5 This is how the Coriolis Effect influences the circulation of the earth's atmosphere. The airflow is deflected to the right in the northern hemisphere, to the left in the southern.

If we apply this deflection to the circulation of the atmosphere as shown in Fig. 1–3, it becomes the more confusing pattern of Fig. 1–5.

The air that flows away from the poles is diverted toward the west and becomes the Polar Easterlies. The flow from 30° to 60° is deflected toward the east and becomes the Prevailing Westerlies.

At the point where the Polar Easterlies and the Prevailing Westerlies converge is a band of low pressure known as the subpolar low.

The flow of air from the Horse Latitudes, a band of high pressure at 30°, to the Doldrums at the equator is diverted to the west and becomes the legendary trade winds.

This, then, is an idealized version of the planetary winds as they would be on a king-size bowling ball. Fortunately for all living things, the earth is made up of land and water, creating an environment conducive to survival. Unfortunately, this same environment upsets the geometric uniformity of the planetary winds.

One reference has been made to the Doldrums out of regard for tradition; in the future it will be referred to as the Intertropical Convergence Zone (ITC). This is, in effect, an undulating meteorological equator where the north and southeast trade winds converge. During the vernal and autumnal equinoxes (March 21 and September 21), it roughly approximates the true equator. In Fig. 1–6 the position of the ITC is shown for July and January. It is radically affected by the distribution of land and water. Since the land absorbs heat more rapidly, the ITC dips to the south over Africa and South America. The oceans are colder in the southern hemisphere, so the ITC is bent to the north in the Atlantic and the eastern Pacific.

Our interest in the position of the ITC lies in the fact that the trade winds move north and south with it.

The Sub Polar Lows that are located near 60° are of lesser interest; only the hardiest sailors frequent those waters. In the southern hemisphere they form a continuous band on the edge of Antarctica while in the northern hemisphere they take the form of several low pressure cells. The most important are the Aleutian Low in the North Pacific and the Iceland Low in the Atlantic. The latter is responsible for the severe gales common to that area during the winter months. In the summer they almost lose their identity.

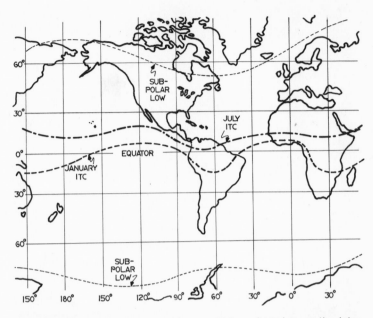

Fig. 1–6 Subsequently the Doldrums will be referred to as the Inter-tropical Convergence Zone (ITC). The low pressure bands at the equator (ITC) and at 60° latitude are not coincidental or even parallel with the lines of latitude; they wander to the north and south in a serpentine fashion and travel north and south with the seasons.

If you are wondering what happened to the Horse Latitudes, take a look at Fig. 1–7. The Horse Latitudes are made up of a series of permanent and semipermanent high pressure cells. The Pacific High and the Bermuda (Azores) High are quite permanent, drifting slightly north, east, south, and west. The continental high is less reliable and is sometimes completely displaced by a low pressure cell.

Now we graduate from the idealized planetary winds to a simplified version of what they are really like (Fig.'s

1–8 and 1–9). These are reasonably accurate, but it should be obvious that you cannot determine what the prevailing winds are on Long Island Sound or Lake Michigan.

The trade winds are remarkably constant, actually blowing in the directions they are supposed to over 80% of the time at an average velocity of force 4. This consistency is because the pressure gradient and temperature are working together to accomplish the same end. The air

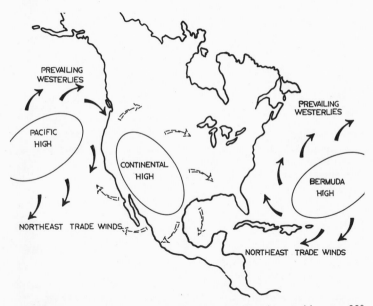

Fig. 1–7 The high pressure band that is supposed to reside near 30° latitude is even more unreliable than the ITC and the subpolar low. Instead of being a continuous band, it turns out that there is a series of permanent and semipermanent high pressure cells, at irregular intervals, wandering well to the north and south of 30°. Fortunately, some of them, like the Bermuda High, are usually just about where they are supposed to be. At least they never disappear.

Fig. 1–8 The typical winter planetary winds show the trade winds and prevailing westerlies at their southernmost limit. The heavy arrows indicate winds of more than 50% constancy and the light arrows those of less than 50% constancy.

is flowing from the high pressure of the Horse Latitudes to the low pressure of the equatorial region, picking up heat as it moves over the warmer water. By the time it reaches the ITC, the air is buoyant and ready to ascend.

It is quite a different matter in the flow of air from the Horse Latitudes to the subpolar low. The pressure gradient causes a flow to higher latitudes, but higher latitudes mean lower surface temperature. The air of the prevailing westerlies gets colder and more dense, which weakens the pressure gradient. No wonder the westerlies are so fickle and have an unreliable constancy quotient of from 10% to less than 50%!

SUMMARY: Much of the preceding is background data that serves no purpose except to clarify some information that is useful. It is important to understand the Coriolis Effect and how it deflects *anything* moving over the earth's surface. It is equally worth remembering what the planetary winds are and their constancy. Determining the

Fig. 1–9 In the summer the whole planetary wind system moves northward, and the trade winds pump warm moist air over the southeastern states, while the west coast is kept cooler by the trade winds that originate over northern land and water. Heavy arrows—constancy, 50% plus; light arrows—less than 50%.

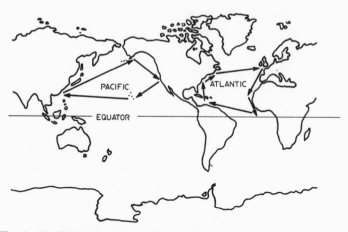

Fig. 1–10 This is how the sailing ships of the eighteenth and nine-teenth centuries took advantage of the planetary winds. In the Atlan-tic, New England skippers sailed for Europe on the prevailing westerlies, rode the back of the Bermuda High to the Gold Coast of Africa, and then the trade winds to the West Indies and home. In the Pacific, they sailed the trade winds to Hawaii and the Orient and the prevailing westerlies back to California.

direction and average velocity of the prevailing winds on the waters upon which you venture is up to you; it is be-yond the scope of this text to catalog the prevailing winds on Chesapeake Bay, Lake Ponchartrain, and Juan de Fuca Strait.

If you remember that the winds are caused by pressure gradients, which in turn are caused by the heating and cooling of the air, you are well on the way to achieving a basic understanding of weather phenomena.

TWO  EPISODIC WINDS

NOW, that we have the planetary winds all straightened out, everything should be quite simple; we know the location and direction of the trade winds and the prevailing westerlies and are ready to face the elements, secure in our expertise.

Not quite—there is some more information about winds that is prerequisite to disappearing hull down over the horizon. But before we get involved in the cause and effects of episodic winds, let's get back to academics again.

It has been pointed out that air flows from areas of high pressure to lows. We all know that the air pressure at sea level is 1013.2 millibars (29.92″ of mercury). Lines drawn through points of equal pressure on a weather map are isobars. A line perpendicular to the isobars is the pressure gradient and its direction is from high to low pressure.

These terms are graphically depicted in Fig. 2–1. When the isobars are close together, the pressure change takes place in less distance than when they are farther apart. When contour lines on a map are closer together

25

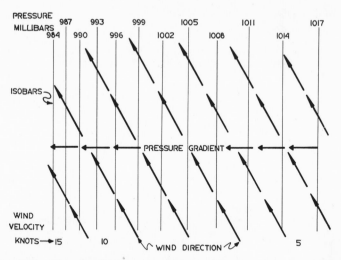

Fig. 2–1 When the isobars are close together the pressure gradient is steep and the wind velocity is higher.

the change in elevation takes place in less distance than when they are farther apart. In the latter case the gradient is said to be steeper. Isobars on a weather map are considered analogous to contour lines on a land map. It is interesting to note that the pressure gradient and the wind direction do not coincide. The wind direction as shown is called the cyclostrophic wind; if under certain conditions the wind should blow parallel to the isobars, it is then known as the geostrophic wind.

In Fig. 2–1 the isobars from 1017 to 984 millibars get closer together as the pressure decreases. This indicates a steeper pressure gradient and an increasing wind velocity. Between the widely spaced 1017 and 1014 millibar isobars the wind velocity is only 5 knots. Where the isobars are closest, between 987 and 984 millibars, the velocity has increased to 15 knots.

This brings the Coriolis Effect back into the picture. When there is an area of high pressure adjacent to a low pressure cell (Fig. 2–2), there is a pressure gradient from high to low and we might expect a flow in that direction. Whoosh! A high and a low become one medium pressure cell. Sorry, but it just doesn't work like that. When the air starts to move, the rotation of the earth deflects it to the right (in the northern hemisphere). Instead of the air rushing directly from the high into the low, it flows outward around the center of the high in a clockwise pinwheel pattern; it is not circular nor is it spiral; it does form spiral bands. The pattern around the center of the low is a series of counterclockwise spiral bands.

It has been previously indicated that the planetary

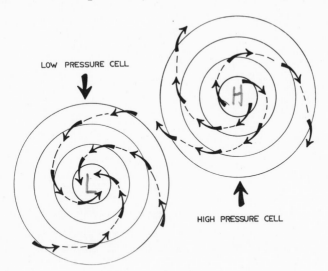

LOW PRESSURE CELL

Fig. 2–2 The wind around a low pressure cell forms a counterclockwise series of spiral bands. Around a high pressure cell the pattern is clockwise.

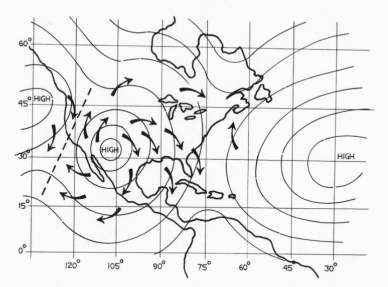

Fig. 2–3 In the temperate zone adjacent high pressure areas are separated by a trough of low pressure indicated by the heavy broken line. This is a condition that may generate an extratropical low.

winds in the temperate zone (prevailing westerlies) are not as reliable as the trade winds of the tropics. A factor that contributes to this situation, in addition to the opposing pressure and temperature gradients, is a series of more or less permanent high pressure areas encircling the earth near 30° latitude. Extending north and south between these highs are troughs of low pressure. And therein lies the trouble.

In Fig. 2–3 there is a high over the southwestern United States and one to the northwest out over the Pacific. The flow of air is clockwise around both in a pinwheel pattern of spiral bands. There is a trough of low pressure indicated by the heavy broken line where the winds converge. Just as wind blowing over the surface of water creates waves, so do opposing winds (Fig. 2–4).

Usually, but not always, a nucleus of low pressure is formed by closed isobars. This is the genesis of an extratropical low pressure system.

Extending to the southwest out of this low is a cold front (triangles indicate cold front and direction of travel). Extending to the northeast is a warm front (semicircles on the side toward which the front is moving). It should be emphasized that extratropical lows are indigenous to the temperate zone and form only between 30° and 60° latitude.

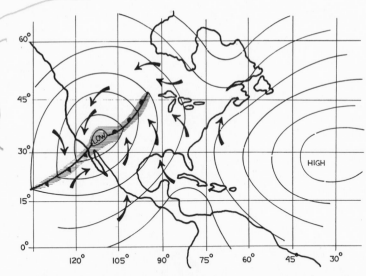

Fig. 2–4 Opposing winds along the low pressure trough create a wave and then a low pressure cell with closed isobars. Here we have the birth of an extratropical low pressure system with a cold front to the southwest and a warm front to the northeast.

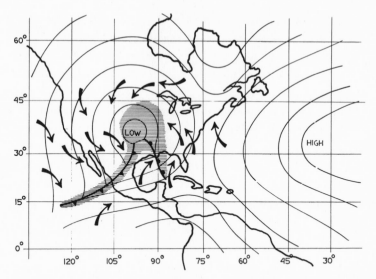

Fig. 2–5   Here is a mature extratropical low with widespread precipitation as indicated by the shaded area. The typical counterclockwise wind pattern is fully developed and has completely replaced the planetary winds.

At this point, our interest in extratropical lows is primarily their effect on the planetary winds. In Fig. 2–3 the wind off Baja California was southeasterly. In Fig. 2–4, after the low has formed the wind has veered to the southwest. In the Gulf of Mexico (Fig. 2–3) the wind was northwesterly; now (Fig. 2–4) it has clocked to the southeast. Around the Great Lakes it was northwesterly, now it is southeasterly.

In Fig. 2–5 the center of the low is over Kansas; the planetary winds off Baja California are back to normal; in the Gulf of Mexico, the wind is now out of the southwest; at the Great Lakes it is still southeasterly.

In another twenty-four hours the low is centered over

Indiana (Fig. 2–6) and the cold front has overridden the warm front, resulting in an occlusion. The Gulf of Mexico has westerly winds and off the east coast from Cape Hatteras to Maine, the wind has backed to the east; at the Great Lakes it is northeasterly; over the Bahamas it has veered to the southwest.

And this is how an extratropical low exerts its influence on the planetary or prevailing winds in the temperate zone. It should be pointed out that the wind shifts which are depicted here are only typical of one path of a frontal system; there can be many variations on the same theme.

When Yankee skippers invade Florida for the Southern

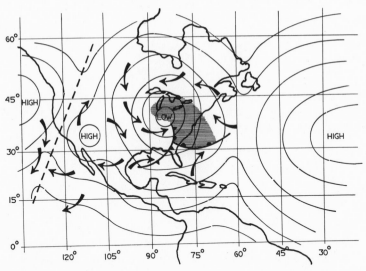

Fig. 2–6 The cold front has overridden the warm front, forming an occluded front. This is the beginning of the end; an extratropical low is about to decay. Out on the Pacific coast there is another low pressure trough between two highs, and the whole process will be repeated.

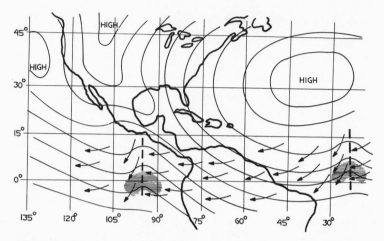

Fig. 2–7 The heavy broken lines northeast of South America and south of Mexico are troughs of low pressure known as Easterly Waves. These tropical disturbances move across the tropics from east to west and are the typical foul weather phenomena in southern waters.

Ocean Racing Conference (S.O.R.C.) circuit, they remark that the weather is different than it is up north. Usually they mean that it is warmer. Regardless of temperature or sky cover, the weather is totally different. Only occasionally does the fringe of the weather pattern associated with extratropical lows invade the Gulf of Mexico and never as far south as the Caribbean.

The only thing that disrupts the reliable trade winds is a series of easterly waves that travel from east to west between latitudes 5° and 15° north in the winter and between 10° and 25° north in the summer.

An easterly wave is a trough of low pressure that appears on the weather map as a wave in the isobars. It creates an area of disturbed weather (Fig. 2–7) shaped like a kidney bean.

In Fig.'s 2–7, 2–8, and 2–9, the effect of an easterly wave on the planetary winds in both the eastern Atlantic and western Pacific is shown.

Fig. 2–9 indicates that the wave has developed into a tropical depression. This distinction is made when there are one or more closed isobars and wind has risen to force 7. The precipitation associated with an easterly wave becomes more intense and widespread.

In either case, the winds associated with one of these tropical disturbances back as the wave approaches and veer as it passes.

Fig.'s 2–3 through 2–9 are idealized versions of the weather patterns that produce the episodic winds. It might be interesting to associate weather conditions as

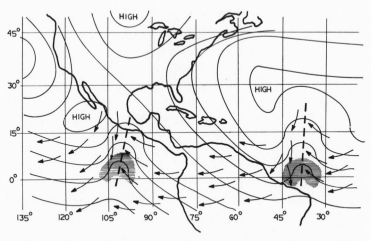

Fig. 2–8 As an easterly wave develops, precipitation becomes widespread (shaded area) and the trade winds of the tropics are replaced by spiral wind bands.

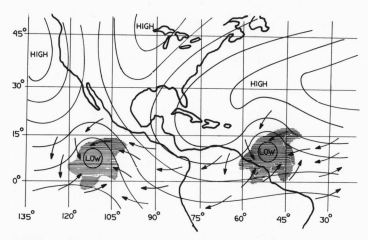

Fig. 2–9 When the spiral wind bands form and winds increase to force 7, the storm is classified as a tropical depression, which may or may not develop into a tropical storm and then a full fledged hurricane.

they actually developed with a specific time, place, and event—for example, the 1972 St. Petersburg-Ft. Lauderdale race, which covered the period of February 5–6–7.

In order to broaden the scope we will review the daily U.S. Weather Bureau maps for the two days preceding and following the race.

The 7:00 A.M. map (Fig. 2–10) showed a weak low pressure cell in the northeastern Gulf of Mexico. A shallow pressure gradient resulted in southeasterly winds of force 2 at St. Petersburg, southwesterly winds of force 2 at Key West, and ESE winds of force 1 at Miami. Since these were early morning land readings, the midday velocity over open water could be expected to about double and clock slightly. Behind a stationary cold front to the west is a real honking Texas Norther; the wind velocity over open water was probably in the force 8 range. If this weather pattern persisted and moved to the east in the next twenty-four hours, the start of the race could have been a real donnybrook.

Well, there goes our donnybrook! The low pressure has moved off over the Atlantic and broken up (Fig. 2–11). The Texas Norther has lost its steep pressure gradient and our winds in the eastern Gulf have picked up only slightly. Force 3 NNE at St. Petersburg, northwesterly force 2 at Miami, and force 3 NNW at Key West.

The race started at noon on Saturday, February 5 (Fig. 2–12); the 7:00 A.M. wind at St. Petersburg was force 2, but by noon the wind over open water was force 4 out of

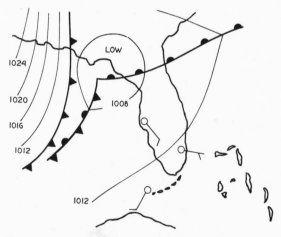

Fig. 2–10 Thursday, February 3, 1972, two days before the St. Petersburg-Ft. Lauderdale Race. The planetary winds are southeasterly and are being reinforced by the pressure gradient caused by the low in the northeastern Gulf of Mexico. A typical Texas Norther with a steep pressure gradient is behind the stationary front extending southwest out of the low pressure cell.

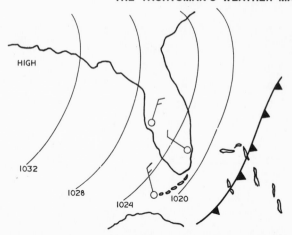

Fig. 2–11 Friday, February 4, 1972. The extratropical low has been filled by the Bermuda High, and the front of the Texas Norther is over the Bahamas. The pressure gradient has so weakened that the northerly winds have diminished to force 3 or less.

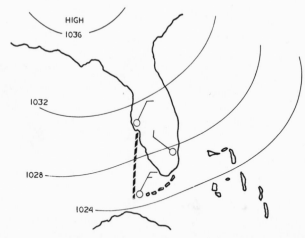

Fig. 2–12 Saturday, February 5, 1972. The first leg of the St. Petersburg-Ft. Lauderdale S.O.R.C. race from Tampa Bay to Rebecca Shoals is shown in the heavy broken line. The weak pressure gradient holds the morning land winds to force 2–3.

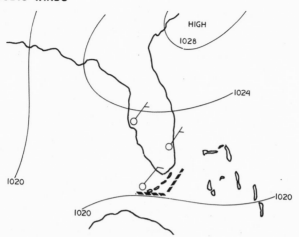

Fig. 2–13 Sunday, February 6, 1972. The second leg of the race through the Florida Straits is provided with northeasterly winds of force 4 onshore at midday.

the northeast; it was a nice spinnaker reach all the way to Rebecca Shoals (west of Key West).

By Sunday (Fig. 2–13) the pressure gradient had really flattened out, and the morning wind was between force 1 and 2. By midmorning the winds over open water were in the force 3 range, and with the help of a 2-knot Gulf Stream the fleet made pretty good time through the Straits of Florida.

Monday, February 7 (Fig. 2–14), showed a cold front moving in over the Florida panhandle, but the wind at the Miami weather station was only force 1 at 7:00 A.M. Offshore it was a little breezier, and with the help of the northbound Stream the fleet moved well.

Practically all the boats had crossed the finish line by 7:00 A.M. Tuesday and a weak cold front had moved out over the Bahamas; the winds were light and northerly

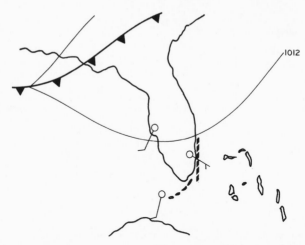

Fig. 2—14 Monday, February 7, 1972. The final leg is sailed under light southeasterly winds, while a cold front approaches from the northwest.

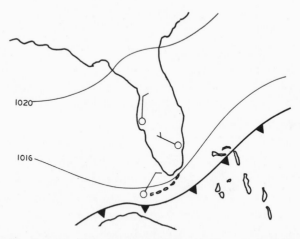

Fig. 2—15 Tuesday, February 8, 1972. By this time most of the fleet has crossed the finish line and so has the cold front. The wind shift was a full 180° but of a very mild nature.

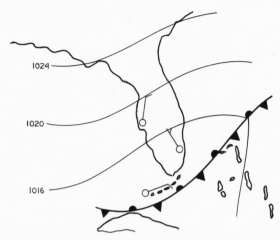

Fig. 2–16 Wednesday, February 9, 1972. The race is all over and the front has pulled up stationary in the Florida Straits. Wind speed is still minimal and another St. Petersburg-Ft. Lauderdale race is sailed in mild weather with light winds.

(Fig. 2–15). Next morning (Fig. 2–16) the front had stalled. If the 180° wind shift at Key West seems unaccountable, note the closed 1016 millibar isobar at the northeast end of the stationary front. This would indicate a low pressure cell somewhere southwest of Cuba.

SUMMARY: The whole key to being weather-wise is a basic understanding of high and low pressure cells and their influence on local conditions. In addition you should also know their location at the time in question in relation to your position. It pays to study the weather maps in the daily newspaper and on TV, even if you are not anticipating the use of your boat. You will soon discover that there is a continuity or sequence to meteorological events which will be applicable at a later time and place.

THREE

LOCAL WINDS, GUSTS, AND EDDIES

WE have observed the effect of temperature and rotation on the planetary winds; we have examined the effect of the episodic winds on the primary circulation; now let us explore the effect of microweather on both.

Most of the winds of microweather are the result of thermal conditions or obstructions to the smooth laminar flow of air. If a small parcel of air is heated at the surface, it will rise aloft. It does not leave a vacuum; cooler, heavier air from above will sink down to replace it. Even if there is no horizontal movement in the air prior to this interchange, there will be when the exchange takes place. As the heated air rises, adjacent surface air will flow in to replace it and only then will the air aloft descend.

If there is horizontal airflow prior to this thermal action and reaction, then surface air moving in to fill the void will reinforce, divert, or reduce its speed and direction.

The term "Sea Breeze" is bandied about in a most familiar fashion. The assumption is made that it is always

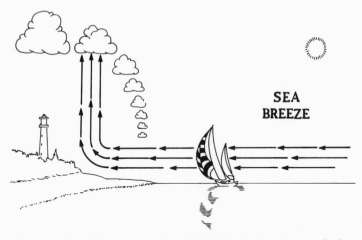

SEA BREEZE

Fig. 3–1 When the air along the shore is heated by the sun, it rises aloft and the air over the water flows in to replace it. Sometimes a row of cumulus clouds will form at the top of the updraft. Occasionally the clouds build sufficiently to become cumulonimbus and late afternoon showers result.

there—a natural feature of the locality. When it fails to appear on schedule, sailors blame it on luck or something equally elusive.

Since the sun heats the land surfaces at a more rapid rate than the water, the air directly above the beaches, shores, or city streets is heated more rapidly than the air over water. Heated air expands and becomes buoyant; it rises aloft, and air from the adjacent body of water flows in to replace it. That's a sea breeze (Fig. 3–1). When you see a line of cumulus clouds lined up above and parallel to a shore line, then you know there's a sea breeze, but you should also know that it is not necessarily perpendicular to the shore.

What happens to our sea breeze when the prevailing

pressure gradient is parallel to the shore, resulting in a wind at right angles to the normal direction of the sea breeze? The combined pressure gradients cause the wind to blow at something between 0° and 90° to the shore, and the velocity will be affected by the angular relationship of the two gradients.

For example, at Cape Hatteras in the summer months, the prevailing (planetary) winds are southerly at force 4 to 6, and no sea breeze is going to change that. But suppose there were a low pressure cell over the Bahamas that had a pressure gradient sufficient to cancel out the southerly flow of air? If it were warm and sunny, Hatteras would have a sea breeze in the middle of the day.

We can have every conceivable combination of prevailing wind, pressure gradient, and local condition to produce a net wind of almost any direction from a gentle breeze to a full gale. The problem is to be able to recognize the causes and to estimate the effects. In the waters that we frequent, we should determine what the prevailing winds are at various times of the year. We should be aware of the presence of high and low pressure cells and the winds associated with them. And above all, we should learn about the microweather that is peculiar to the area.

If you don't know what the prevailing winds are, ask someone who does; Fig. 1–8 and Fig. 1–9, which are superficial in scope, will give you an approximation. Study the weather map in your daily newspaper or on TV. Better still, cut the weather map out of the paper and keep a daily tabulation of the wind direction. You'll soon find that a definite pattern becomes apparent over a period of time. If the prevailing wind is only 10% constant, the

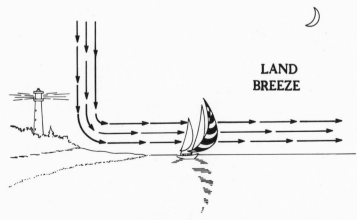

Fig. 3–2 The heated air that ascended during the day is cooled after sundown and becomes heavy enough to sink to the surface and flow out over the water.

other 90% will be scattered all over the compass rose.

This same weather map analysis will produce the knowledge that, for example, there is an extratropical low approaching from the west and will apparently pass to the north of your position, that your prevailing wind will shift first to the southeast and then to the northwest as the cold front passes you.

Now, if in addition to this, you have also made a study of the effect of local terrain on the winds, then you are well on your way to developing the understanding of wind and weather that is so vital to the skilled sailor.

There is a corollary to the sea breeze—the land breeze. It takes place after sundown when the heated air that has risen along the shore line is cooled and descends, flowing out over the water (Fig. 3–2). It is less pronounced than the sea breeze and is also affected by the prevailing and gradient winds.

The land breeze or night wind is a cool and shallow zephyr that is easily diverted or stalled by obstructions. It is fine for running out to sea or for pussyfooting along the shore. For tacking against a foul tide, it is next to useless except in deep valleyed estuaries where it is accelerated and reinforced by gravity (Fig. 3–6).

Suppose the shoreline, instead of being a gentle slope, is a palisade or steep bluff? Then what happens to the sea or land breeze? For that matter, what happens to a regular onshore or offshore wind? If the wind is blowing onshore (Fig. 3–3), it piles up at the foot of the bluff creating a miniature high pressure cell that is almost devoid of wind. What exists is fluky and full of eddies; the direction is as much vertical as it is horizontal. It's a bad place to get caught under sail.

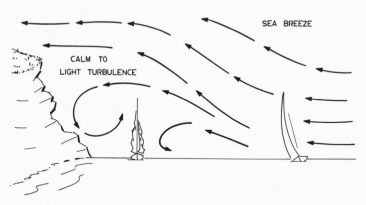

Fig. 3–3 When there is an onshore breeze against a steep shore line, a pocket of eddies and turbulence forms. In light airs it is a dangerous spot for sailboats without power. In strong winds it is unsafe for any kind of craft.

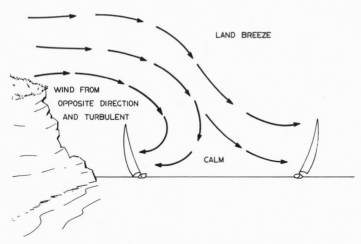

Fig. 3–4 When the wind blows out over a bluff, it causes a reverse flow of air at the base and an area of calm a little offshore. The original airstream does not come to the surface of the water until it has traveled several times the height of the promontory.

An offshore wind (Fig. 3–4) goes scooting out to a distance several times the height of the promontory. The lower stratas of the laminar flow descend to the surface of the water and flow out and in toward shore. Where the airflow is vertical there is an area of calm; close to shore, the direction is toward the rocks and disaster.

Let us take a look at another type of terrain. Many rivers and streams that flow into a larger body of water wend their way through a valley or canyon. During the day the air in the valley is heated and it rises, creating a low pressure along the valley floor. The result is an up-valley wind (Fig. 3–5). If the valley is steep and deep enough, the sun may heat one slope before the other, creating some interesting variations. It would pay to study

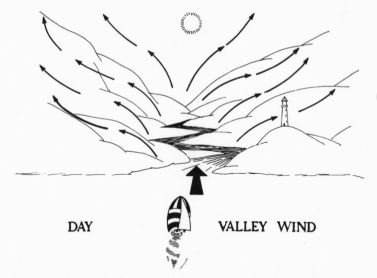

DAY   VALLEY WIND

Fig. 3–5 During the heat of the day the air in a valley is heated and it ascends. Due to the mechanical efficiency of funnellike terrain, the daytime valley wind is much more powerful than the corresponding type sea breeze.

the characteristics of the actual valley winds in your area.

At night the wind direction is reversed and the cooled air descends down the slopes, building up volume and speed (Fig. 3–6). When it reaches the mouth of the estuary it may build up to healthy proportions. Witness the mistral of the Riviera, favored by mystery writers. It is a valley wind that plunges down from the Alps into the Rhone Valley and out over the Gulf of Genoa. This icy blast puts a devastating chill on the playground of the jet set and produces gale force winds for Mediterranean sailors.

The understanding of the relationship between surface winds and winds aloft separates the men from the boys when it comes to racing. Let's broaden that to include winds over land areas. If the observed wind at a shore-based weather station indicates that the wind is NNE, the wind offshore will be clocked a few degrees in relation to that. The wind aloft will clock a few degrees in relation to the surface wind over open water. If the wind ashore is 10 knots, the wind offshore will be on the order of 15 knots and the wind aloft will be close to 20 knots. All these figures are relative and merely used to illustrate the point.

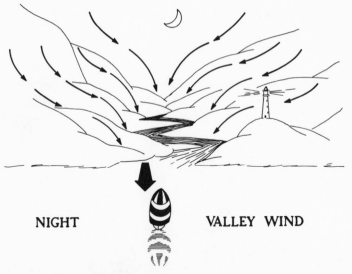

NIGHT   VALLEY WIND

Fig. 3–6   Like the day valley wind, the night valley wind is one of the most powerful of all thermal air currents. It is the only one in which a yacht can tack with any degree of success against a foul current.

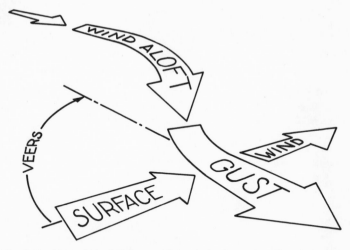

Fig. 3–7 The wind aloft has a higher velocity and a direction somewhat clockwise from the surface wind. When, because of cooling, it descends to the surface, it does so as a gust of wind veered from the surface wind.

In Fig. 3–7 are illustrated the relationship of the wind aloft to the surface wind and the manner in which it descends to the surface, creating a gust that veers from the surface wind. When you are sailing in fluky and gusty air, it pays to remember that the gusts will veer and have a higher velocity than the existing surface winds.

Fig. 3–8 shows how this can effect two boats, one on the port tack and one on the starboard. When the gust hits them, the port tack boat gets a header, has to fall off and come about. The boat on the starboard tack gets favored, heads up slightly, and picks up a lead of at least a boat length.

There are enough variations on the theme of local

winds, gusts, and eddies to fill a book (it has filled several), but the best place to learn about them is out on the water. Armed with a basic understanding of the principles involved, experience and observation are the best teachers.

SUMMARY: Living testimonial to the last paragraph is Ron Krippendorf, first winner of the National Miniature Ocean Racing Club (M.O.R.C.) Championship. Ron has an

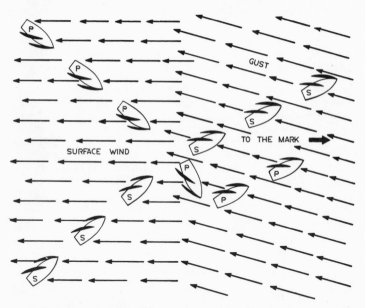

Fig. 3–8 The yacht (P) is on the port tack and sailing a converging course with yacht (S). When the wind gusts, S is favored and simply heads up slightly. P gets a header and has to fall off and then come about on the starboard tack, over a boat length behind. So having the right-of-way is not the only advantage of being on the starboard tack.

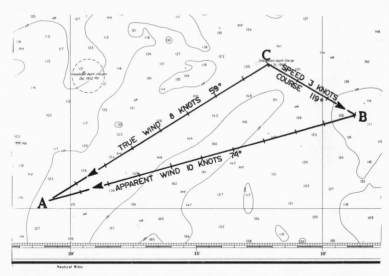

Fig. 3–9 Here is a simple way to determine the velocity and direction of the true wind when the apparent wind, true course, and boat speed are known. On the chart lay out the yacht's course and speed to any convenient scale (C–B). From point B, lay out the apparent wind to the same scale. Line A–C is the direction and speed of the true wind. For the benefit of anyone who doesn't know what the apparent wind is: a boat moving through the water creates a wind of its own, equal to its forward speed. The combined result of this and the true wind is called the apparent wind. The true wind is always farther aft than the apparent wind. If the apparent wind is dead astern, the true wind is either dead ahead or dead astern.

insatiable thirst for meteorological knowledge, keeps notebooks, records wind, tide, and current data on old charts, and makes extensive use of radio and a battery-operated TV. It should be emphasized that Ron doesn't collect information; he accumulates knowledge.

FOUR

TIDES AND CURRENTS

THERE is an irrevocable bond between wind and water. The planetary winds are a major motivating force in those rivers of the seas, the ocean currents; the episodic winds can lash the tranquil waters into a tempest of towering waves.

The seas heat and cool the air; the moisture content of the atmosphere is a gift of the oceans of the world. No study of marine weather is complete without at least a superficial analysis of tides and currents.

The gravitational pull of the moon and the sun move whole oceans, not to mention the earth itself and its atmosphere. Even humans are affected to the extent that they gain and lose minutely with the tides.

The moon's gravitational pull causes the seas to bulge in the direction of the lunar attraction and also result in a bulge on the opposite side of the earth because centrifugal force overcomes the moon's influence (Fig. 4–1).

Although the sun's pull is much less, it alternately reduces or reinforces the moon's pull. When the sun and moon are in line, they join forces to create an unusually

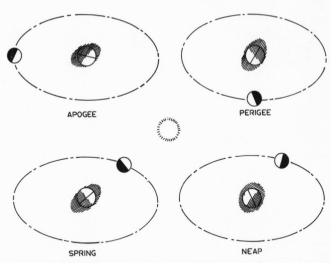

Fig. 4–1 These are a few of the forces that influence the rise and fall of the tides. The moon follows an elliptical path about the earth. When it is at its most distant point (apogee), the tides are lower than when the moon is closest (perigee) to the earth. When the moon and sun are aligned, the tides are at their highest (spring). When the moon is perpendicular to the sun, the tides are extremely low (neap). Add to this the centrifugal force of the earth's rotation and you have an infinite variation of forces affecting the tidal fluctuation.

high, high (spring) tide, and when the sun and moon are perpendicular to each other, the sun cancels out some of the moon's pull and the result is an exceptionally low high (neap) tide.

Because the tidal flow is not confined to a uniformly symmetric basin, the results are complex and varied. In mid-ocean, where there is little to restrain the tidal flow, the tide is little more than a foot or so. In the Bay of Fundy, between Nova Scotia and New Brunswick, the tide rises more than forty feet (Fig. 4–3). If you remember the way the tide comes in at the beach, creeping over

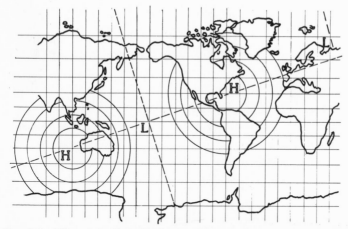

Fig. 4-2 When it is high tide in the Bahamas, it is also high tide at a point approximately opposite, somewhere west of Australia. There is an irregular band of low tide encircling the earth and roughly perpendicular to the tidal axis.

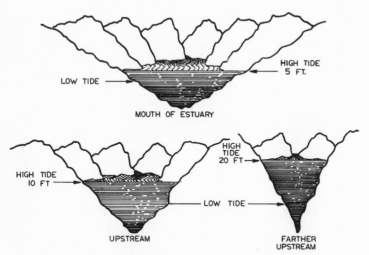

Fig. 4-3 When the incoming tide is forced into a deep, narrow passage, the tide is progressively higher the farther it goes upstream. This sometimes results in a tidal bore—a wall of water several feet high boiling through an estuary like a wave breaking on a beach.

the sand several feet for each inch of vertical rise, you will be able to visualize that when the flow is confined in a narrow valley with steep shores the rise would be much greater.

The whole technical exposition of tides is a complicated and involved science and is far beyond the scope of this text. It will suffice most marine interests to know when the tide will be high or low and to what degree. This information is available in the Tide Tables published by the National Oceanic and Atmospheric Administration.

Just as the atmosphere is in constant motion, so are the waters of the oceans. The surf pounding the beaches of California may have spent a century creeping along the ocean floor from Antarctica.

The similarity of the seas to the atmosphere is more than mere motion; one is almost a mirror image of the other. At the equator the water is heated by the sun's rays and expands; the sea level there is several inches higher than normal. Since water naturally flows downhill, the equatorial surface waters flow north and south. To compensate for the displacement, the cold polar waters flow toward the equator along the ocean floor.

The eastward rotation of the earth has a tendency to leave the seas behind to pile up on the western shores. Evidence of this is the fact that the sea level on the Atlantic side of the Isthmus of Panama is several feet higher than on the Pacific side. Then that old Coriolis Effect reappears and the ocean currents are deflected to the right north of the equator and to the left, south.

When we impose the planetary wind system on top of

this flow, it is reinforced and accelerated. Another factor involved in the path of the ocean currents is that of continental deflection. Obviously, when a large land mass obstructs the flow of an ocean current, its path is diverted to a new direction. The Yucatan Peninsula, for example, turns the western set of the North Equatorial Current northward into the Gulf of Mexico.

In Fig. 4–5 we see an idealized version of the ocean currents around North America. Between the North and South Equatorial currents is a counter current which is

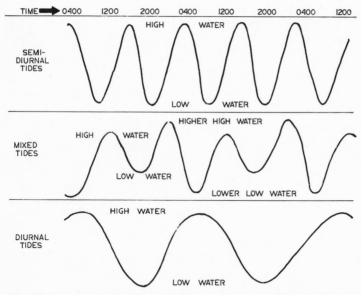

Fig. 4–4 It is not surprising that the tidal characteristics vary from time to time and from place to place. The so-called normal tide is semidiurnal when there is a high tide every 12 hours and 25 minutes. There is also a mixed tide that has high tide and higher high tide, low tide and lower low tide. Then there is the diurnal tide which has only one high and one low per lunar day (24 hours and 50 minutes).

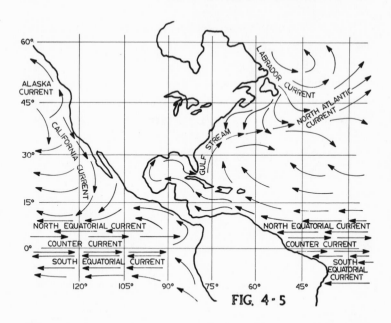

FIG. 4 - 5

the result of water piling up on the coast of South America augmented by the low atmospheric pressure along the Intertropical Convergence zone (ITC) where the winds are virtually calm.

Limiting our survey to those currents which occur in the northern hemisphere adjacent to the coasts of North America, the most important are the North Equatorial currents. In the Atlantic the set is almost due west with a drift that ranges from 0.6 knot near the Cape Verde Islands to over one knot in the Caribbean. The Equatorial Currents are the most powerful and encompass the largest volume of water; they are the source of all other currents.

The North Equatorial Current piles up against the Yucatan Peninsula, reinforced with a component from the South Equatorial Current. It pours into the Gulf of Mexico, where part of it turns to the east and flows out through the Florida Straits. The remainder circulates in the Gulf of Mexico and then joins the eastward set along the Keys. In the southern Bahamas this flow merges with another component of the North Equatorial Current to form the most famous of all the rivers of the sea—the Gulf Stream.

This remarkable current is noted for its constancy and velocity, which reaches its peak drift of almost 3 knots from the tip of Florida to north of the Bahamas. The drift gradually decreases to about one knot at Cape Hatteras, where the Gulf Stream turns northeastward and heads for Europe. East of Newfoundland it splits into the two components of the North Atlantic Drift. One continues past Iceland and Norway, the other continues to the east and curves southward off the coast of Europe, where it becomes the Canaries Current and rejoins the North Equatorial Current at the Cape Verde Islands.

In the icy Davis Strait between Labrador and Greenland, we find a portion of the Gulf Stream flowing northward along the west coast of Greenland, curving around and following the Labrador shores from which the Labrador Current gets its name. It takes a southerly set to the Grand Banks, where it is split into a current that flows down the eastern coast of the United States and one that joins the Gulf Stream on its trans-Atlantic trek. Another portion sinks deep into the sea and slowly returns to the equatorial waters from which it came.

The union of the Gulf Stream and the Labrador Current is the spawning ground for the infamous fogs of the Grand Banks. The Labrador Current is also responsible for the coastal fogs all the way to Florida.

Since the waters west of Central America are the beginning of the Pacific North Equatorial Current, there is little direct connection between them and the currents that prevail along the western coast of North America. The slow drifting Alaska Current and California Current move at speeds of less than one knot. Their influence is mostly meteorological. They bring fog to the entire west coast north of the 30th parallel.

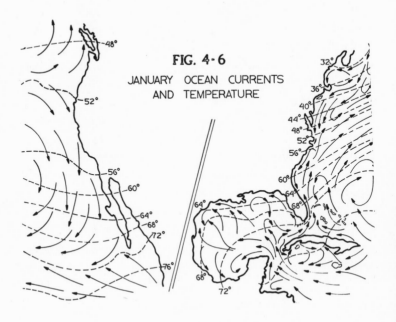

FIG. 4-6

JANUARY OCEAN CURRENTS
AND TEMPERATURE

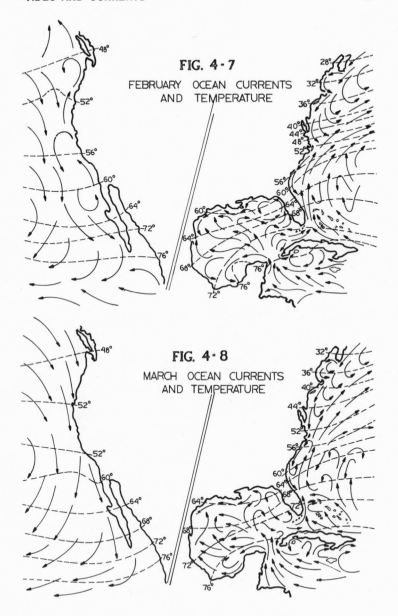

FIG. 4 - 7

FEBRUARY OCEAN CURRENTS
AND TEMPERATURE

FIG. 4 · 8

MARCH OCEAN CURRENTS
AND TEMPERATURE

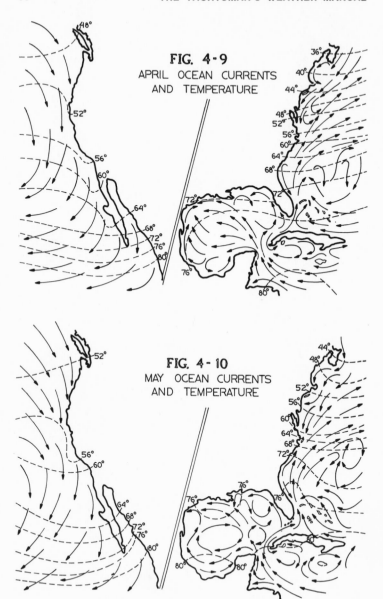

FIG. 4-9
APRIL OCEAN CURRENTS
AND TEMPERATURE

FIG. 4-10
MAY OCEAN CURRENTS
AND TEMPERATURE

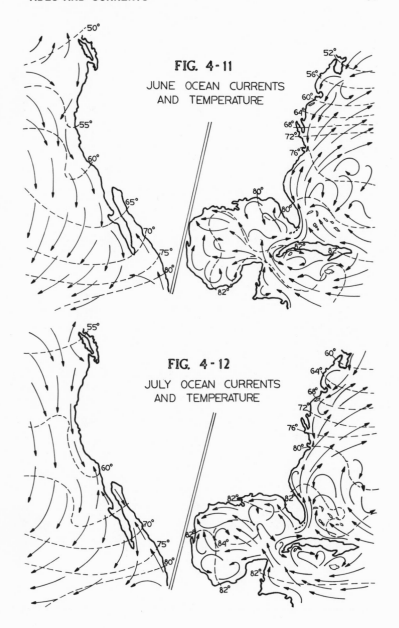

FIG. 4-11

JUNE OCEAN CURRENTS
AND TEMPERATURE

FIG. 4-12

JULY OCEAN CURRENTS
AND TEMPERATURE

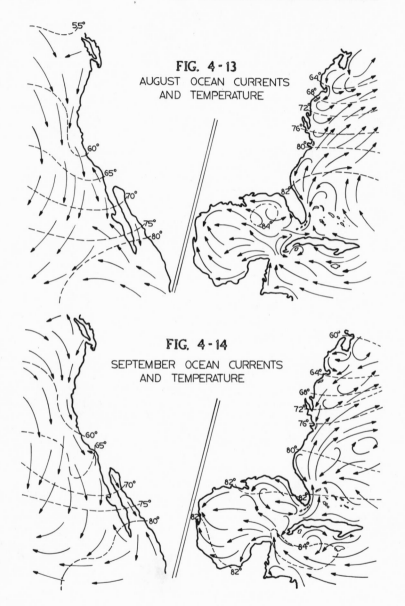

FIG. 4 - 13
AUGUST OCEAN CURRENTS
AND TEMPERATURE

FIG. 4 - 14
SEPTEMBER OCEAN CURRENTS
AND TEMPERATURE

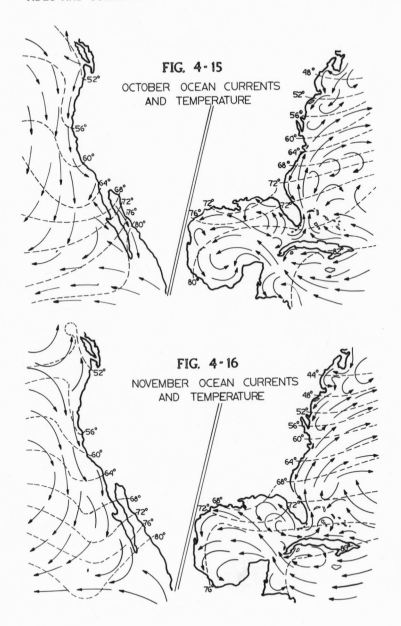

FIG. 4-15

OCTOBER OCEAN CURRENTS
AND TEMPERATURE

FIG. 4-16

NOVEMBER OCEAN CURRENTS
AND TEMPERATURE

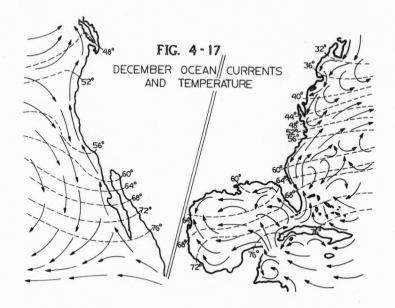

FIG. 4-17
DECEMBER OCEAN CURRENTS
AND TEMPERATURE

SUMMARY: The essence of this chapter is depicted in Fig. 4–5 through Fig. 4–17. A careful study of these will reveal a wealth of information relative to the weather and yachting conditions along the coastal waters on both sides of the continent of North America. Current arrows indicate averages. The actual currents encountered may differ; they may even be opposite in direction.

FIVE

THE BATTLE
OF THE GIANTS:
EXTRATROPICAL LOWS

IDEOLOGICAL wars are waged between nations with different economic and political philosophies.

Meteorological wars are waged between continent-sized air masses with different temperature and humidity characteristics.

When a large parcel of air comes to rest or slowly moves over a surface of fairly uniform temperature and moisture, then the air will tend to become similar to the underlying surface.

An air mass is an enormous high pressure cell which tends to remain homogenous; it has surface tension similar to a drop of water on an oily surface and spreads out only when its volume and internal pressure increase, causing the air mass to bulge in the direction of lower pressure.

This tendency for stability provides the time factor required for the air mass to acquire the characteristics of the surface upon which it lies. If it forms over cold, dry land, it becomes a cold, dry air mass; if it forms over warm water, it becomes a warm, humid air mass.

They are named according to their source: Polar, Arctic, Tropic, or Equatorial. The distinctions made by Arctic and Equatorial are so academic they will be disregarded here.

The subcategories that relate to source are designated by "m" for maritime and "c" for continental. If the air mass is colder than the surface over which it is moving, the letter "k" is added. If it is warmer, the letter "w" is added.

cPk = continental Polar air, cooler than the surface over which it is moving.

mTw = maritime Tropic air, warmer than the surface over which it is moving.

In Alaska, northern Canada, Baffin Island, Lands End, and Victoria Island, continental Polar air (cPk) collects. It attacks the United States during winter months along a broad front (Fig. 5–1). The eastern path of cPk brings sub-zero weather and when it collides with maritime Tropic air (mTw), violent winter storms result. As it moves over the warmer waters of the Atlantic or the Great Lakes, fog develops.

The central path of cPk frequently charges down the Mississippi Valley, bringing winter at its worst to the Gulf of Mexico. Sometimes it is blocked by a task force of mTw air, in which case a winter storm system is formed to rush off to the northeast and out into the North Atlantic.

APPENDIX

CONVERSION TABLE—INCHES OF MERCURY TO MILLIBARS

Inches	Millibars	Inches	Millibars	Inches	Millibars	Inches	Millibars
27.50	931.3	27.94	946.2	28.38	961.1	28.82	976.0
27.51	931.6	27.95	946.5	28.39	961.4	28.83	976.3
27.52	931.9	27.96	946.8	28.40	961.7	28.84	976.6
27.53	932.3	27.97	947.2	28.41	962.1	28.85	977.0
27.54	932.6	27.98	947.5	28.42	962.4	28.86	977.3
27.55	933.0	27.99	947.9	28.43	962.8	28.87	977.7
27.56	933.3	28.00	948.2	28.44	963.1	28.88	978.0
27.57	933.6	28.01	948.5	28.45	963.4	28.89	978.3
27.58	934.0	28.02	948.9	28.46	963.8	28.90	978.7
27.59	934.3	28.03	949.2	28.47	964.1	28.91	979.0
27.60	934.6	28.04	949.5	28.48	964.4	28.92	979.3
27.61	935.0	28.05	949.9	28.49	964.8	28.93	979.7
27.62	935.3	28.06	950.2	28.50	965.1	28.94	980.0
27.63	935.7	28.07	950.6	28.51	965.5	28.95	980.4
27.64	936.0	28.08	950.9	28.52	965.8	28.96	980.7
27.65	936.3	28.09	951.2	28.53	966.1	28.97	981.0
27.66	936.7	28.10	951.6	28.54	966.5 ·	28.98	981.4
27.67	937.0	28.11	951.9	28.55	966.8	28.99	981.7
27.68	937.4	28.12	952.3	28.56	967.2	29.00	982.1
27.69	937.7	28.13	952.6	28.57	967.5	29.01	982.4
27.70	938.0	28.14	952.9	28.58	967.8	29.02	982.7
27.71	938.4	28.15	953.3	28.59	968.2	29.03	983.1
27.72	938.7	28.16	953.6	28.60	968.5	29.04	983.4
27.73	939.0	28.17	953.9	28.61	968.8	29.05	983.7
27.74	939.4	28.18	954.3	28.62	969.2	29.06	984.1
27.75	939.7	28.19	954.6	28.63	969.5	29.07	984.4

CONVERSION TABLE—INCHES OF MERCURY TO MILLIBARS

Inches	Millibars	Inches	Millibars	Inches	Millibars	Inches	Millibars
27.76	940.1	28.20	955.0	28.64	969.9	29.08	984.8
27.77	940.4	28.21	955.3	28.65	970.2	29.09	985.1
27.78	940.7	28.22	955.6	28.66	970.5	29.10	985.4
27.79	941.1	28.23	956.0	28.67	970.9	29.11	985.8
27.80	941.4	28.24	956.3	28.68	971.2	29.12	986.1
27.81	941.8	28.25	956.7	28.69	971.6	29.13	986.5
27.82	942.1	28.26	957.0	28.70	971.9	29.14	986.8
27.83	942.4	28.27	957.3	28.71	972.2	29.15	987.1
27.84	942.8	28.28	957.7	28.72	972.6	29.16	987.5
27.85	943.1	28.29	958.0	28.73	972.9	29.17	987.8
27.86	943.4	28.30	958.3	28.74	973.2	29.18	988.2
27.87	943.8	28.31	958.7	28.75	973.6	29.19	988.5
27.88	944.1	28.32	959.0	28.76	973.9	29.20	988.8
27.89	944.5	28.33	959.4	28.77	974.3	29.21	989.2
27.90	944.8	28.34	959.7	28.78	974.6	29.22	989.5
27.91	945.1	28.35	960.0	28.79	974.9	29.23	989.8
27.92	945.5	28.36	960.4	28.80	975.3	29.24	990.2
27.93	945.8	28.37	960.7	28.81	975.6	29.25	990.5
29.26	990.9	29.70	1,005.8	30.14	1,020.7	30.58	1,035.6
29.27	991.2	29.71	1,006.1	30.15	1,021.0	30.59	1,035.9
29.28	991.5	29.72	1,006.4	30.16	1,021.3	30.60	1,036.2
29.29	991.9	29.73	1,006.8	30.17	1,021.7	30.61	1,036.6
29.30	992.2	29.74	1,007.1	30.18	1,022.0	30.62	1,036.9
29.31	992.6	29.75	1,007.5	30.19	1,022.4	30.63	1,037.3
29.32	992.9	29.76	1,007.8	30.20	1,022.7	30.64	1,037.6
29.33	993.2	29.77	1,008.1	30.21	1,023.0	30.65	1,037.9
29.34	993.6	29.78	1,008.5	30.22	1,023.4	30.66	1,038.3
29.35	993.9	29.79	1,008.8	30.23	1,023.7	30.67	1,038.6
29.36	994.2	29.80	1,009.1	30.24	1,024.0	30.68	1,038.9
29.37	994.6	29.81	1,009.5	30.25	1,024.4	30.69	1,039.3
29.38	994.9	29.82	1,009.8	30.26	1,024.7	30.70	1,039.6
29.39	995.3	29.83	1,010.2	30.27	1,025.1	30.71	1,040.0
29.40	995.6	29.84	1,010.5	30.28	1,025.4	30.72	1,040.3
29.41	995.9	29.85	1,010.8	30.29	1,025.7	30.73	1,040.6
29.42	996.3	29.86	1,011.2	30.30	1,026.1	30.74	1,041.0
29.43	996.6	29.87	1,011.5	30.31	1,026.4	30.75	1,041.3
29.44	997.0	29.88	1,011.9	30.32	1,026.8	30.76	1,041.7
29.45	997.3	29.89	1,012.2	30.33	1,027.1	30.77	1,042.0
29.46	997.6	29.90	1,012.5	30.34	1,027.4	30.78	1,042.3
29.47	998.0	29.91	1,012.9	30.35	1,027.8	30.79	1,042.7
29.48	998.3	29.92	1,013.2	30.36	1,028.1	30.80	1,043.0

CONVERSION TABLE—INCHES OF MERCURY TO MILLIBARS

Inches	Millibars	Inches	Millibars	Inches	Millibars	Inches	Millibars
29.49	998.6	29.93	1,013.5	30.37	1,028.4	30.81	1,043.3
29.50	999.0	29.94	1,013.9	30.38	1,028.8	30.82	1,043.7
29.51	999.3	29.95	1,014.2	30.39	1,029.1	30.83	1,044.0
29.52	999.7	29.96	1,014.6	30.40	1,029.5	30.84	1,044.4
29.53	1,000.0	29.97	1,014.9	30.41	1,029.8	30.85	1,044.7
29.54	1,000.3	29.98	1,015.2	30.42	1,030.1	30.86	1,045.0
29.55	1,000.7	29.99	1,015.6	30.43	1,030.5	30.87	1,045.4
29.56	1,001.0	30.00	1,015.9	30.44	1,030.8	30.88	1,045.7
29.57	1,001.4	30.01	1,016.3	30.45	1,031.2	30.89	1,046.1
29.58	1,001.7	30.02	1,016.6	30.46	1,031.5	30.90	1,046.4
29.59	1,002.0	30.03	1,016.9	30.47	1,031.8	30.91	1,046.7
29.60	1,002.4	30.04	1,017.3	30.48	1,032.2	30.92	1,047.1
29.61	1,002.7	30.05	1,017.6	30.49	1,032.5	30.93	1,047.4
29.62	1,003.1	30.06	1,018.0	30.50	1,032.9	30.94	1,047.8
29.63	1,003.4	30.07	1,018.3	30.51	1,033.2	30.95	1,048.1
29.64	1,003.7	30.08	1,018.6	30.52	1,033.5	30.96	1,048.4
29.65	1,004.1	30.09	1,019.0	30.53	1,033.9	30.97	1,048.8
29.66	1,004.4	30.10	1,019.3	30.54	1,034.2	30.98	1,049.1
29.67	1,004.7	30.11	1,019.6	30.55	1,034.5	30.99	1,049.5
29.68	1,005.1	30.12	1,020.0	30.56	1,034.9		
29.69	1,005.4	30.13	1,020.3	30.57	1,035.2		

NOTE: 1 inch = 33.86395 millibars; 1 millibar = 0.02952993 inch; 1 millimeter = 0.039370 inch; 1 inch = 25.4005 millimeters; 1 millibar = 0.7500616 millimeter; 1 millimeter = 1.33322387 millibars.

BIBLIOGRAPHY

Coles, K. Adlard: *Heavy Weather Sailing*. Tuckahoe, New York: John De Graff, Inc., 1968.

Donn, William L.: *Meteorology with Marine Applications*. New York: McGraw-Hill, 1951.

Dunn & Miller: *Atlantic Hurricanes*. Louisiana State University Press, 1960.

Miller, Albert: *Meteorology*. Columbus, Ohio: Charles E. Merrill Publishing Co., 1966.

Murchie, Guy: *Song of the Sky*. Boston: Houghton Mifflin, 1954.

Stewart, George R.: *Storm*. New York: Random House, 1941.

Sutton, O. G.: *The Challenge of the Atmosphere*. New York: Harper & Bros., 1961.

SOURCE MATERIAL

Daily Weather Maps, National Oceanic and Atmospheric Administration

Pilot Chart of the North Pacific, N.O. 55

Pilot Chart of the North Atlantic, N.O. 16

U.S. Naval Oceanographic Office

GLOSSARY

MANY of these terms are beyond the technical scope of this book; they are listed here for the benefit of those who might be inclined to pursue the subject in greater detail.

Absorption—The ability to receive and store heat by radiation.

Adiabatic—The gain or loss in heat without heat transfer. A gas cools when it expands and becomes warmer when compressed.

Advection—Heat transfer by the horizontal movement of air. Advection fog forms when warm moist air is cooled as it moves over cold water.

Advisories—Severe weather warnings issued by the Weather Bureau. See Gale and Hurricane advisories.

Air mass—A large parcel of air, hundreds of miles in diameter and five to ten miles high. The temperature and humidity of an air mass is homogeneous and is acquired from the surface over which it forms.

Air mass analysis—The use of the characteristics and movement of air masses to forecast the weather.

Aleutian Low—A semipermanent low in the North Pacific; during the winter, it is relatively warm; during the summer, it is relatively cold and has little effect on the weather.

Altocumulus—Middle cloud of fleecy globules; also called mackerel sky.

Altostratus—Middle cloud, sheetlike, grayish or bluish in color and usually covering most or all of the sky.

Anemometer—A device for measuring wind velocity.

Aneroid barometer—An instrument for measuring atmospheric pressure.

Antarctic Low—A belt of low pressure at about 60° south latitude.

Anticyclogenesis—The formation of a new anticyclone or the intensification of an existing one.

Anticyclone—A high pressure cell with its typical clockwise wind pattern (see cyclone).

Anvil top—The elongated formation at the top of a cumulonimbus cloud created by the winds aloft.

Apparent wind—The observed wind on a boat under way; it is the resultant of the true wind and the ship's wind (boat speed).

Arctic air mass—A high pressure cell that forms within the Arctic Circle and acquires the temperature and humidity of the surface.

Arctic Low—A belt of low pressure at about 60° north latitude.

Arctic sea smoke—Advection fog that forms when extremely cold air moves over warmer water.

Atmospheric envelope—The layer of gases that enshrouds the earth.

Azores High—A permanent high pressure cell that is part of the Horse Latitudes between the 30th and 40th parallels and centered near the Azores (also called the Bermuda High).

Backing—A windshift in a counterclockwise direction (west to southwest, northeast to north).

Barograph—A registering or recording barometer.

Barometer—An instrument for measuring atmospheric pressure.

Beaufort Scale—A numerical value given to wind velocity based on visual observations (see text).

Bermuda High—See Azores High.

Boyle's Law—At a constant temperature, the volume of a gas varies inversely with the pressure on it. Thus if we double the pressure of a gas, the volume decreases by one half, and conversely.

Breakers—Sea waves that become too high for stability; the crest breaks and rolls.

Buys-Ballot's Law—When facing the wind in the northern hemisphere, the low pressure is to the observer's right, and conversely.

cAk—Continental Arctic air, cooler than the surface over which it is moving.

Cat's paws—Individual wavelets caused by a light, fitful wind.

Centrifugal force—The inertial reaction by which a body tends to move away from the center about which it is revolving.

Charles' Law—At a constant pressure, the volume of a gas varies directly with the temperature, and conversely.

Chinook—A warm, dry wind descending the leeward side of a mountain; also called a foehn.

Chubasco—A violent thundersquall along the west coast of Central America.

Cirrocumulus—Mackerel sky—a high form of cumulus clouds.

Cirrostratus—A high overcast cloud formation through which the sun or moon shines faintly, sometimes with a halo.

Cirrus—A high form of wispy clouds called mares' tails; they are made up of ice crystals.

Clocking—Veering—a clockwise windshift.

Col—An area of moderate pressure with two opposing highs and two opposing lows on its perimeter.

Cold front—The leading edge of a cold air mass overtaking a warm air mass.

Condensation nuclei—Minute particulants of solid matter about which water vapor collects to form raindrops.

Conduction—Heat transfer by direct contact without motion.

Convection—Air in vertical motion, either up or down.

Convergence—The flow of air in toward the center of a low pressure cell. The meeting of winds (the northeast and southeast Trade Winds converge near the equator).

Cordonazo—A violent wind blowing from the southern quadrant as a result of a hurricane passing off the coast of Mexico.

Coriolis Effect—The rotation of the earth causes the path of a body moving over the earth's surface to be deflected to the right in the northern hemisphere and to the left in the southern. It is maximum at the poles and zero at the equator.

cPk—Continental Polar air, cooler than the surface over which it is moving.

Cumulonimbus—Low based clouds of great vertical development caused by convection. There is usually precipitation and gusty winds.

Cumulus—Low level clouds caused by thermal convection. They are fleecy and cottonlike in appearance.

Cyclogenesis—The process by which a low pressure (cyclone)

grows and develops into a storm system, either tropical or extra-tropical.

Cycloidal waves—These are seas of recent origin; they are short and choppy with pronounced peaks or crests.

Cyclone—This term has been studiously avoided in the text because of its frequent misuse. Its precise meaning is any low pressure cell with its accompanying counterclockwise wind system. It does not imply a storm or bad weather of any kind. Unfortunately, it is the name given to East Indian hurricanes. It is also used interchangeably with "tornado." In either case, it is technically incorrect (see anticyclone).

Cyclostrophic wind—The Coriolis Effect, pressure gradient, and centrifugal force result in a wind direction aloft parallel to the isobars. At the surface, friction causes the wind to cross the isobars at a slight angle; thus is caused the cyclostrophic wind.

Deepening—Cyclogenesis.

Depression—A low pressure cell; there must be at least one closed isobar.

Dew—The condensation of water vapor on a smooth surface as a result of cooling the air below its dew point.

Dew point—The temperature at which air becomes saturated and its water vapor condenses.

Divergence—Descending air reaches the surface and flows out in a clockwise pattern; basically a high pressure cell.

Doldrums—A belt of calm airs encircling the earth near the equator. See Intertropical Convergence Zone.

Easterly Wave—An area of low pressure moving from east to west in the Trade Winds (see text).

Eddy—A deflected air current that reverses direction and flows in a more or less circular path.

Entrainment—The mixing of environmental air with convective currents. Downdrafts and updrafts entrain the air through which they are passing.

Equatorial Front—See Intertropical Convergence Zone.

Evaporation—The transformation of water or water drops into water vapor. The process consumes heat and lowers the ambient temperature.

Extratropical cyclone—Extratropical low pressure system (see text).

Eye of a hurricane—The central core of minimum pressure and relative calm.

Fetch—The distance a sea wave has traveled since it was generated by the wind.

Filling—The process of airflow into a low pressure cell and increasing the barometric pressure—frontolysis.

Foehn—The European equivalent of Chinook.

Fog—A cloud at zero altitude.

Dense fog	no visibility at	50	yards
Thick fog	" "	200	"
Fog	" "	500	"
Moderate fog	" "	1000	"
Thin fog	" "	2000	"

Frontogenesis—The intensification and development of an extratropical low pressure system.

Frontolysis—The weakening and dissolution of an extratropical low pressure system.

Fronts—The boundary line between air masses of different temperature and humidity.

Gale—Wind of force 8—34 to 40 knots.

Gale warning—Issued to the public and maritime interests twenty-four hours before winds of force 8 are expected either over the coast or the waters adjacent.

General circulation—The basic airflow of the earth's atmosphere; the planetary winds.

Geostrophic wind—When the wind blows parallel to the isobars.

Gradient wind—Airflow generated by pressure differential.

Halo—A luminous ring around the sun or moon as their rays are refracted by the ice crystals or cirrostratus clouds.

Humidity—Water vapor suspended in air.

Absolute—The actual weight of the water vapor contained in a specific volume of air.

Specific—The actual weight of water vapor contained in a specific weight of air.

Relative—The percentage of water vapor in a parcel of air with saturated air as 100%.

Hurricane—A tropical storm with force 12 winds—64 knots or higher.

Hurricane Warning—Issued to the public and maritime interests twenty-four hours before winds of force 10 are expected either over the coast or the waters adjacent.

Hurricane Watch—Issued to warn the public and maritime interests that high tides or winds may endanger the area within thirty-six hours.

Insolation—Heating by the rays of the sun.

Instability—A condition where light warm air is below dense cold air.

Intertropical Convergence Zone—A band encircling the earth where the northeast and southeast Trade Winds converge.

Inversion—When the temperature increases with altitude instead of decreasing.

Isallobar—Lines on a weather map through points of equal pressure change.

Isobar—Lines on a weather map through points of equal atmospheric pressure.

Isotherm—Lines on a weather map through points of equal temperature.

Katabatic—The gravitational flow of heavy air down a mountain slope.

Lapse rate—The decrease in temperature with the increase in altitude.

Latent heat—The heat used or released in a thermodynamic process.
  —of condensation—released to the atmosphere.
  —of evaporation—taken from the atmosphere.
  —of sublimation—released to or taken from the atmosphere.
  In each case the water temperature remains the same; only the air temperature varies.

Levanter—A strong easterly wind in the Mediterranean accompanied by fog.

Mackerel sky—High flying cirrocumulus or altocumulus clouds.

mAk—Maritime Arctic air, cooler than the surface over which it is moving.

Mares' tails—Cirrus clouds with wispy ends.

Meridional flow—The north and south movement of air.

Micrometeorology—The study of weather phenomena peculiar to a particular area.

Millibar—A unit of atmospheric pressure (see text).

Mistral—A stormy, cold, northerly wind descending from the Alps into the Rhone Valley and the Gulf of Lyons.

Monsoon—Land and sea winds of the Indian Ocean that change seasonally rather than daily.

mPk—Maritime Polar air, cooler than the surface over which it is moving.

mTw—Maritime Tropic air, warmer than the surface over which it is moving.

Nimbostratus—A low sheetlike cloud, frequently with precipitation.

Northeast trade winds—The planetary winds that prevail in the tropical zone of the northern hemisphere.

Norther—A cold winter wind along the coast of the Gulf of Mexico (see text).

Occluded front (occlusion)—The condition that exists when a cold front overtakes a warm front.

Pacific High—A permanent high pressure cell that forms between the coast of California and Hawaii.

Papagayo—A strong northeast wind blowing over the Gulf of Papagayo (Costa Rica) during the winter. Similar to a Texas Norther.

Polar easterlies—The planetary winds that prevail near the 60th parallel.

Polar front—The boundary line between Polar air masses and warmer air masses in the mid-latitudes.

Polar Highs—High pressure cells that build up in the polar regions.

Pressure trough—An area of low pressure with no closed isobars.

Prevailing westerlies—The planetary winds that prevail in the mid-latitudes.

Radiational cooling—Loss of heat into space by the surface and the air in contact with it. It occurs at night under clear skies.

Recurvature—The turning of a tropical storm from the west to the north and east.

Refraction of waves—The change of direction and intensity of sea waves as they approach a shore line.

Ridge of high pressure—An area of high pressure with no closed isobars.

Roll cloud—The leading, low level edge of a cumulonimbus cloud in which the most violent gusts and turbulence occur.

Santa Ana wind—A strong, dry wind that descends from the mountains near Santa Ana, California.

Sargasso Sea—A large area of calm water that approximately corresponds to the Azores-Bermuda High. It is the central core about which the North Atlantic currents rotate.

Saturation—When the air contains all the water vapor of which it is capable, it is saturated—100% relative humidity.

Sea breeze—An onshore wind caused by thermal updrafts along the shore (see text).

Secondary circulation—Surface wind system that results from high and low pressure distribution—the Episodic Winds (see text).

Seiche—The oscillation of an inland body of water caused by winds, pressure variation, earthquakes, etc.

Shear line—The horizontal boundary of two countercurrents of air, usually defined by a layer of high altitude clouds.

Southeast trade winds—The planetary winds that prevail in the tropical zone of the southern hemisphere.

Spiral band—A narrow spiraling band of clouds and rain terminating at the eye of a hurricane or tropical storm. It is in these bands that convergence and rainfall are concentrated.

Stability—A condition of the atmosphere when the air at the surface is cooler than the air aloft; there are no convective currents.

Stratocumulus—A low level, sheetlike cloud of some vertical development.

Stratus—A low level, sheetlike cloud referred to as a low overcast.

Sublimation—The direct transition of ice or snow to water vapor or vice versa with no intermediate liquid stage; a form of evaporation and condensation.

Subsidence—The slow settling or descent of an air mass; accompanied by divergence.

Surge—A wave or rapidly rising water level along a coastline as a result of low pressure and high winds; associated with high winds and other severe storms.

Tehuantepecer—A strong wind on the Pacific coast of Nicaragua that is the result of a high pressure cell over central U.S. A corollary of a Texas Norther.

Temperate Zones—The area between the Tropic of Cancer and the Arctic Circle and the Tropic of Capricorn and the Antarctic Circle.

Tornado—A large whirling funnel of air extending earthward from the base of a cumulonimbus cloud. The cause is thought to be a rapidly moving cold air mass overriding a warm air mass resulting in extreme instability.

Trochoidal waves—Commonly called swells; these are waves that were generated by a distant wind.

Tropical depression—A low pressure cell in the tropics with one or more closed isobars and a wind of force 7.

Tropical disturbance—A low pressure trough in the tropics with no closed isobars.

Tropical storm—A low pressure cell in the tropics with several closed isobars and a wind of force 8.

Tropopause—The boundary between the troposphere and the stratosphere.

Troposphere—That part of the earth's atmosphere in which all convection takes place that results in our weather variables.

True wind—The wind direction and velocity in relation to the surface; it is the observed wind when a boat is stationary in the water.

Turbulence—The combined effects of convection in a specific area.

Typhoon—The name given to hurricanes in the western Pacific.

Veering—When the wind shifts in a clockwise manner (southeast to south, northwest to north).

Vorticity—The tendency of air to spin or rotate as a result of a wind shear or curved wind path.

Warm front—The boundary between a warm air mass overtaking a cold air mass.

Warm sector—In an extratropical low pressure system, the zone between the cold front and the warm front.

Waterspout—The equivalent of a tornado over water.

Water vapor—Invisible water particles suspended in air.

Willie Willie—The Australian term for hurricanes.

Windshift line—The veering of the wind along either a warm or cold front.

Zonal flow—The movement of air east or west.

INDEX

6 - 8
13 - 24
32
39
41
48
54 - 8
65 - 6
73 - 82
83 - 91
92 - 107
108
156 - 8 - 9
163 - 4 - 5
170

Pilot Chart No16